Christmas /97

To: Daniel
with
Love
FROM: Aunt Meena
& Uncle Tommy

READER'S DIGEST Kids®

BIG BOOK OF
SPACE

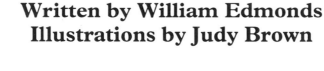

Written by William Edmonds
Illustrations by Judy Brown

A Reader's Digest Kids Book
Published by The Reader's Digest Association, Inc.

Produced by Marshall Editions

Library of Congress Cataloging in Publication Data
Edmonds, William.
  Big book of space / written by William Edmonds : illustrations by
Judy Brown.
      p.   cm.
  Includes index.
  ISBN 0-89577-648-0
  1. Space and time—Juvenile literature. 2. Space perception—
Juvenile literature.   [1. Space and time. 2. Space perception.]
I. Brown, Judy, ill. II. Title.
  Q163. E36 1995
  114—dc20.   94-28252
                                               CIP
                                               AC

2 4 6 8 10 9 7 5 3 1

Editors: Antony Mason, Kate Phelps
Designer: Shaun Barlow
Cover Design: Mike Harnden

Marshall Editions would like to thank Nigel Hawkes,
Science Editor of *The Times*, London, for acting as
consultant on this book.

Printed in Italy

CONTENTS

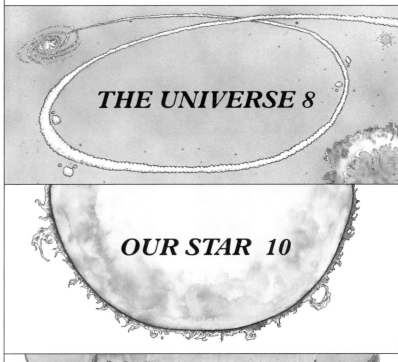

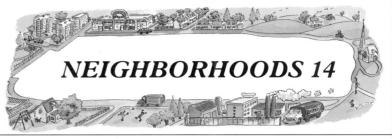
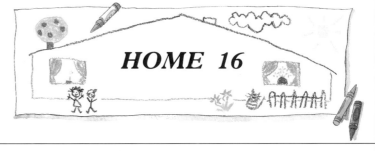
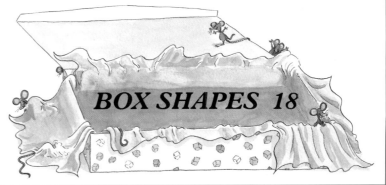

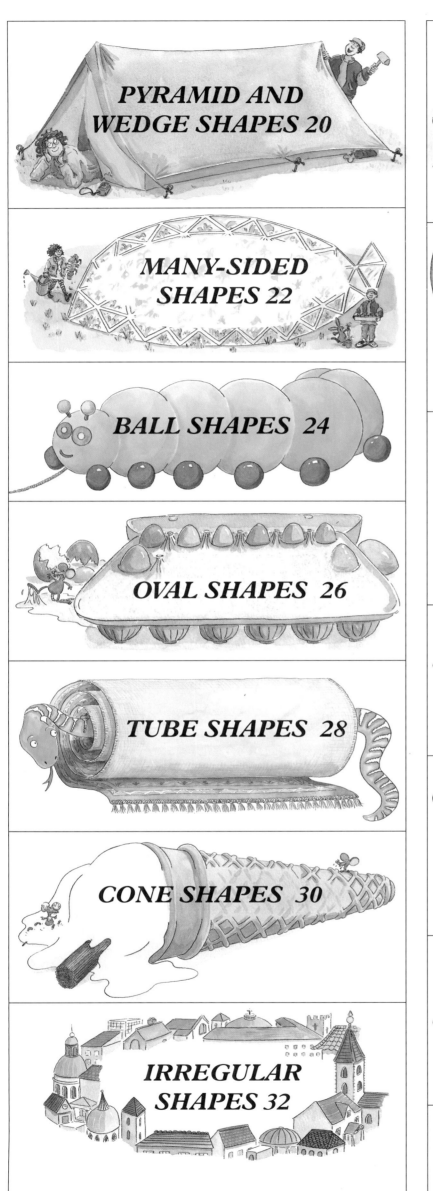

WHAT IS SPACE?

Space is the shape, size, and substance of everything—and of everything in between. It's here, there, and everywhere, where everything happens now, has happened in the past, and will ever happen in the future. It extends in every direction, as far as we can go, see, or even imagine. It's as small as the tiniest speck and as immense as the universe.

Space has many hidden mysteries. Is it emptiness, or is it fullness? Does it change? Is it always there?

You are about to start a journey of adventure, in search of all spaces and of all space.

How small is it?
Start by finding some small spaces, maybe even the space of your own room or smaller spaces within it. Look through a microscope, and you can see the tiniest spaces.

▼ Have a good look around you! Notice how each space has its own shape, and how spaces fit together. Look at a map. It shows you a small chart of the many large spaces there are to explore!

▼ Outside, spaces of all sorts extend far into the distance. If you use a telescope you can see even farther, all the way into outer space!

▼ To learn more about spaces and space, you'll have to go out and about—into your neighborhood, even around the world. And don't forget to stop and look up at the sky—especially at night. What kind of spaces are up there?

How big is it?
Just how far do all the spaces in space stretch? There is so much to find out! For example, how far away are the most distant parts of the world? How far away is the moon? How much space is there between the moon and the stars?

Are you ready?
Prepare yourself for a mission of excitement and discovery! Get ready for the greatest space journey ever! Then climb aboard the space machine and fasten your seat belts!
    Off we go . . . into space!

THE UNIVERSE

The universe is the space that contains all the spaces we know or can imagine. We know of nothing bigger than this space. In it we find ourselves, our world, other planets, and all the stars—galaxies upon galaxies of them.

We call the space beyond our own world and its atmosphere "outer space." It is an enormous emptiness patterned with sprinklings of stars, just a few of which we see at night.

Everything in the universe is constantly moving. Also, we believe that the universe is expanding— the galaxies appear to us to be moving farther and farther away from each other.

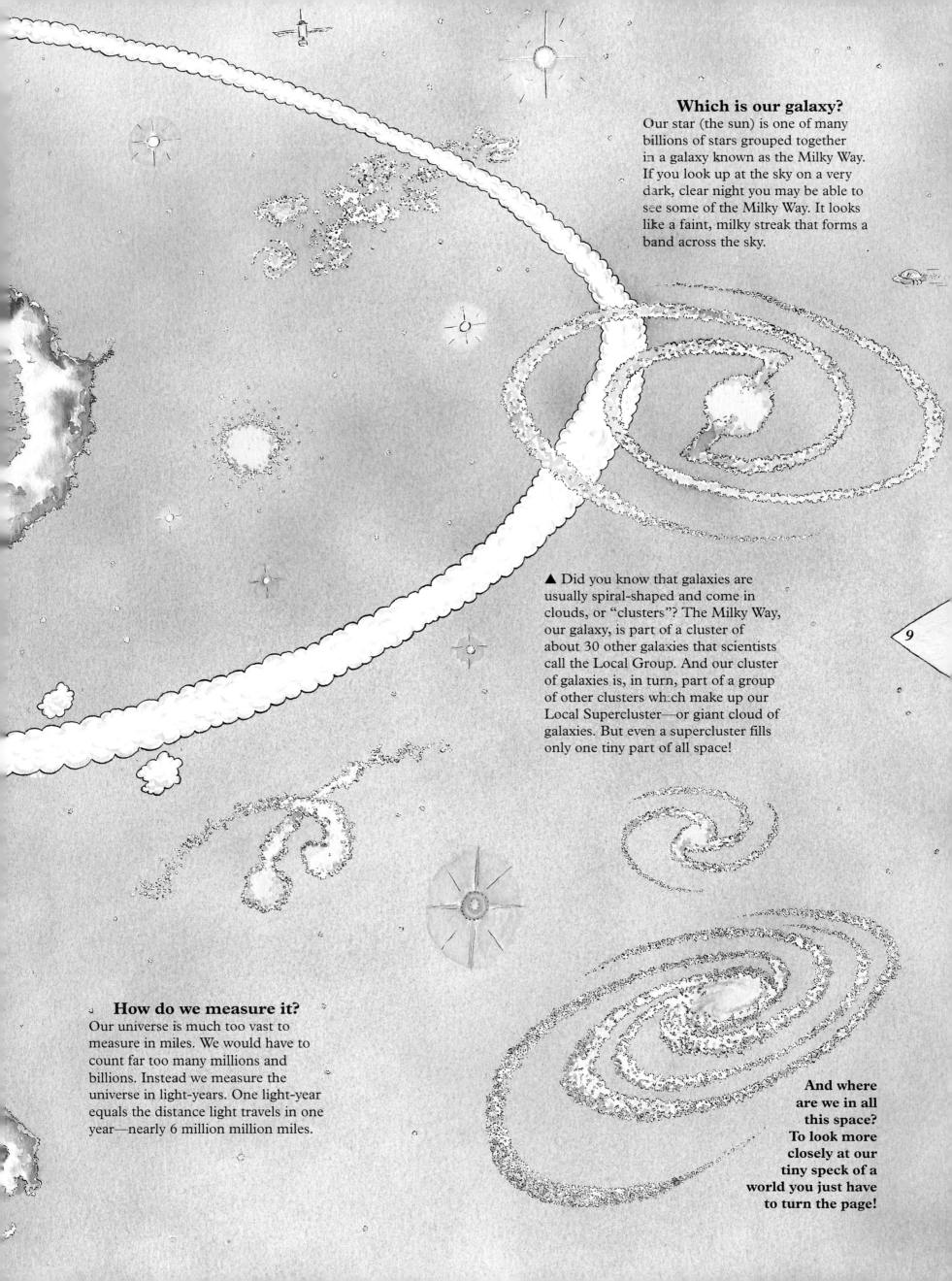

Which is our galaxy?

Our star (the sun) is one of many billions of stars grouped together in a galaxy known as the Milky Way. If you look up at the sky on a very dark, clear night you may be able to see some of the Milky Way. It looks like a faint, milky streak that forms a band across the sky.

▲ Did you know that galaxies are usually spiral-shaped and come in clouds, or "clusters"? The Milky Way, our galaxy, is part of a cluster of about 30 other galaxies that scientists call the Local Group. And our cluster of galaxies is, in turn, part of a group of other clusters which make up our Local Supercluster—or giant cloud of galaxies. But even a supercluster fills only one tiny part of all space!

9

How do we measure it?

Our universe is much too vast to measure in miles. We would have to count far too many millions and billions. Instead we measure the universe in light-years. One light-year equals the distance light travels in one year—nearly 6 million million miles.

And where are we in all this space? To look more closely at our tiny speck of a world you just have to turn the page!

OUR STAR

At the center of our tiny part of the universe is that great, glorious, golden fireball, the sun— our very own star. It looks quite small to us, but that's because it is so far away. In fact, it occupies a million times more space than the Earth.

Nine planets orbit (travel around) the sun. A ring of rocks called asteroids also orbits the sun. Together all these orbits create a dishlike space that we call the solar system. This is our space in the universe, the part that we know most about.

10

Mercury

Moon   Earth

Neptune

Saturn

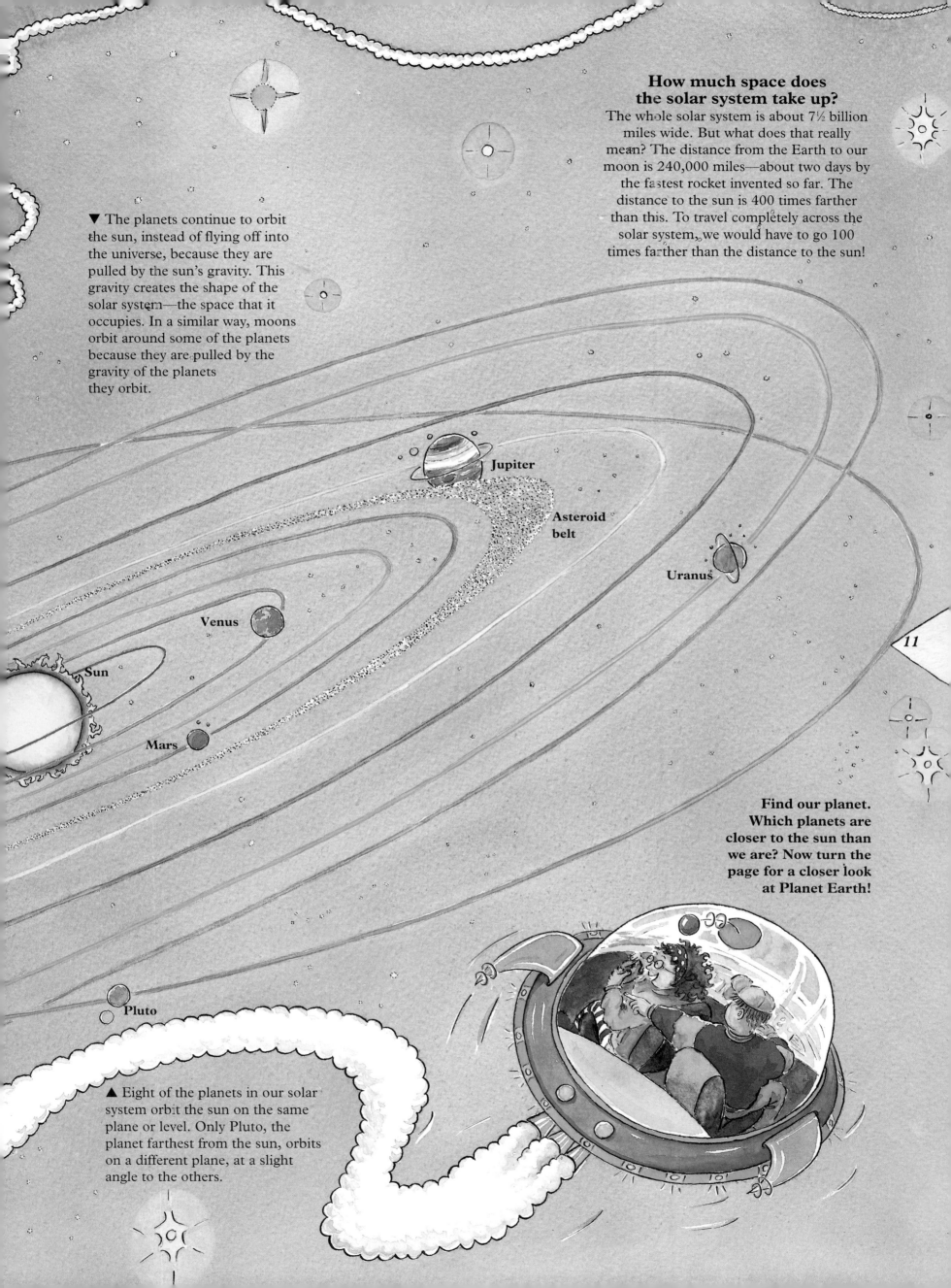

The whole solar system is about 7½ billion miles wide. But what does that really mean? The distance from the Earth to our moon is 240,000 miles—about two days by the fastest rocket invented so far. The distance to the sun is 400 times farther than this. To travel completely across the solar system, we would have to go 100 times farther than the distance to the sun!

▼ The planets continue to orbit the sun, instead of flying off into the universe, because they are pulled by the sun's gravity. This gravity creates the shape of the solar system—the space that it occupies. In a similar way, moons orbit around some of the planets because they are pulled by the gravity of the planets they orbit.

Jupiter

Asteroid belt

Uranus

Venus

Sun

Mars

11

Find our planet. Which planets are closer to the sun than we are? Now turn the page for a closer look at Planet Earth!

Pluto

▲ Eight of the planets in our solar system orbit the sun on the same plane or level. Only Pluto, the planet farthest from the sun, orbits on a different plane, at a slight angle to the others.

OUR WORLD

crust

mantle

outer core

inner core

12

Our world—our very own space in the solar system—is Planet Earth. This wonderful bit of space has so many exciting parts to explore: hot spaces and freezing cold ones, empty spaces and crowded ones, wide open spaces and dark forested ones.

Throughout history, people have explored more and more of the world's spaces. Now there are few places that people have not visited. But one vast space still remains almost unexplored—the Earth's secret center.

▲ No one knows for certain what fills the space in the middle of the Earth, under the crust on which we live. The center of the Earth, known as the core, is mostly made of dense, hot iron. The inner core is solid but the outer core is liquid. The core is covered by the mantle, a thick, rocky layer. Over the mantle is a thin crust of lighter rock.

▲ What covers most of the world's surface? Believe it or not, it's water! In fact, scientists have measured twice as much water as land on the surface of the Earth.

Bump! Down to Earth we come! Now follow the road and find the spaces you would most like to explore—mountains, deserts, icy seas, jungles, busy cities . . .

Going far?

Travel to the world's most faraway spaces. Cross the vast, hot, dry deserts, or the icy wastes of the polar regions. For miles and miles and miles you may see nothing growing and meet no other human beings.

Going up?

Climb to some of the world's highest spaces, up to the top of snow-capped peaks. You will find that the air there is thin, which may make it hard to breathe. But from such heights you can look out on the open spaces all around.

13

Going to town?

Look for the world's most crowded spaces in the famous big cities, such as New York, São Paulo, and Tokyo. There you will find a mass of people, traffic, buildings, and objects of all kinds packed into every bit of the city's space.

Going down?

To reach the world's lowest spaces you will have to dive to the bottom of the ocean. You will need special equipment to allow you to breathe and to protect you from the great pressure of deep water.

SPACE TO LIVE IN
NEIGHBORHOODS

Neighborhoods are the special spaces where people choose to live in groups. A good neighborhood creates a friendly space, where people know and help each other. It's a small part of the big world that we can really get to know. It's where we live, go to school, go to work, do our shopping, meet our friends, and have fun.

Some neighborhoods, such as those in cities, can be very crowded and tight for space. Other neighborhoods may be very spread out. Telephones, cars, and airplanes make connections between neighborhoods. As a result, people living on the other side of the world may seem like our neighbors!

14

◄ In some remote parts of Australia, a person's nearest neighbors may be hundreds of miles away. School children do their lessons at home. They use a two-way radio to speak to their teachers. That is the closest thing to a neighborhood they have!

▲ When people use radio or telephone to talk to each other between one continent and another, the world seems smaller. It can almost appear to be one single neighborhood—one "global village."

Follow the road into this busy neighborhood. What do you see? Find as many places as you can that look like those in your neighborhood.

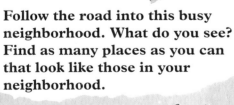

Spaces to choose

If you could choose any neighborhood in the world, where would you live? Would you choose a city space or a country space? Would you live close to the stores, sports clubs, and theaters—or far away from it all?

In the city

One city can contain many neighborhoods. If you want to find extra space in a busy city center, you may have to go UP! That's why city buildings are often so tall. The more crowded a city is, the taller its buildings will probably be.

In a suburb

You can find quieter spaces to live just outside a busy town or city center. In these neighborhoods, called suburbs, houses usually have more space between them, with yards and trees around them.

In a town

A town serves as the center for people who live in the wider open spaces around it. People come to town to buy the things that they cannot find in small, local stores. Towns may also have theaters, banks, and government offices.

In a village

A village is a tiny town, a small community of people living together in their own special space. Some villages are located in remote faraway places, nestling among mountains or in the middle of vast plains. In spaces where most other people are difficult to reach, good neighbors really count.

On the farm

Many farms take up plenty of space. Farmers, their families, and their animals live far away from any neighbors. With so much space to take care of, their life on the farm can be hard work—but lots of fun, too!

SPACE OF OUR OWN
HOME

Welcome home, to your very own space, where you and your family and your things belong. Here you can sleep, cook and eat, wash, help out with the daily chores, or relax and just be yourself.

Homes come in many different shapes and styles, built from all sorts of materials. A few homes, such as palaces, have lots of room and many rooms. Most homes are smaller. Some may be apartments in buildings that rise high into the sky. Others are houses built together in a long, connected row. Still others stand alone on their own piece of land.

Even animals have homes! Can you think of some?

Houses in Yemen

A fishing village in Malaysia

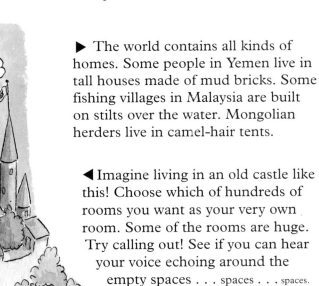

▶ The world contains all kinds of homes. Some people in Yemen live in tall houses made of mud bricks. Some fishing villages in Malaysia are built on stilts over the water. Mongolian herders live in camel-hair tents.

◀ Imagine living in an old castle like this! Choose which of hundreds of rooms you want as your very own room. Some of the rooms are huge. Try calling out! See if you can hear your voice echoing around the empty spaces . . . spaces . . . spaces.

Tents in Mongolia

Look out! All sorts of homes are built along this road. Who do you think has the best kind of space to live in?

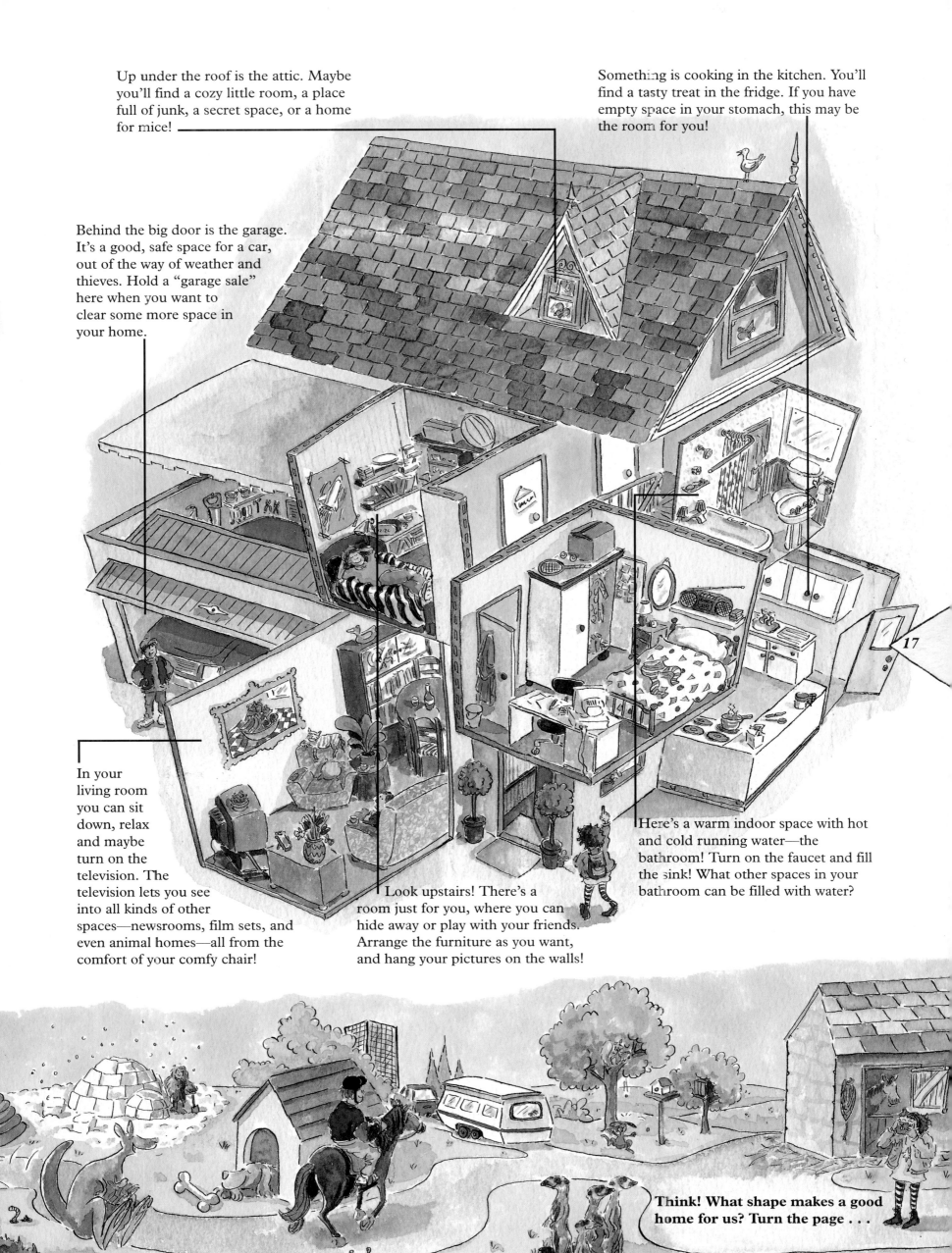

Up under the roof is the attic. Maybe you'll find a cozy little room, a place full of junk, a secret space, or a home for mice!

Something is cooking in the kitchen. You'll find a tasty treat in the fridge. If you have empty space in your stomach, this may be the room for you!

Behind the big door is the garage. It's a good, safe space for a car, out of the way of weather and thieves. Hold a "garage sale" here when you want to clear some more space in your home.

In your living room you can sit down, relax and maybe turn on the television. The television lets you see into all kinds of other spaces—newsrooms, film sets, and even animal homes—all from the comfort of your comfy chair!

Look upstairs! There's a room just for you, where you can hide away or play with your friends. Arrange the furniture as you want, and hang your pictures on the walls!

Here's a warm indoor space with hot and cold running water—the bathroom! Turn on the faucet and fill the sink! What other spaces in your bathroom can be filled with water?

17

Think! What shape makes a good home for us? Turn the page . . .

BOX SHAPES

Now let's get into shape! Let's get straight with space! Boxes are shaped spaces with six straight sides. They come in handy because they fit and stack so well together. They make useful storage spaces that are also convenient to ship.

Boxes with all square sides are known as cubes. Boxes with rectangular sides are known as cuboids. All cubes and cuboids have six sides, eight corners, and twelve edges. Boxes also clearly show us the three dimensions of space: width, height, and depth.

18

▶ You can build with box-shaped blocks. Place one on top of another to make a wall, a house, or even a soaring tower. If you build a set of four walls, you can look inside a whole new straightened space!

Pack and stack as you head along the road. How many of these box-shaped things could you pack in your box-shaped van?

▼ What is the biggest box shape you can think of? Some of the world's largest boxes are "blocks" of apartments or offices. They provide space for many people to live and work closely together. Every one of these buildings is made up of hundreds of smaller box shapes—layer on layer of boxes, and more boxes set side by side.

▲ Can you work out how many little cubes make up each of these bigger cubes and cuboids? You can measure amounts of space by multiplying the measurements of width, height, and depth. For example: 2 inches (width) times 2 inches (height) times 2 inches (depth) equals 8 cubic inches.

Boxes and boxes!
Look for the big cardboard boxes that are used in supermarkets. What's inside them? More boxes! You'll find boxes of breakfast cereal, chocolates, tissues, tea, margarine, cookies, soap powder, cake mix, and much more. Some people think we use too many boxes. Can you imagine a way to use fewer boxes?

▼ To draw accurate box shapes, you need to draw in perspective (so that things in the picture seem to get smaller the farther away they are). You draw the edges of the boxes so they look as though they go toward the distance, as if their lines would all eventually join at one point.

PYRAMID AND WEDGE SHAPES

Pyramids are the shapes made famous by the ancient Egyptians when they built massive pointed tombs for their kings. Egyptian pyramids have four triangular sides with a square base, but pyramids can have three, four, or any number of triangular sides. Wedges are also made with triangles, one at each end with three rectangular sides in between. These pointed shapes are used to make strong spaces such as the roofs of buildings or the pyramids, which have been standing for over 4,000 years!

20

▶ The largest Egyptian pyramid was built as a tomb for King Khufu in about 2500 B.C. Most of the space inside is solid rock—two million huge blocks of cut stone. Far inside the pyramid are the secret spaces of the burial chamber.

◀ Cut a cube in half from corner to corner and what do you make? Two wedges. These shapes are also known as prisms.

Here's a winding road! Find all the pointed shapes along the way!

◀ See how useful wedge shapes can be! A ramp helps the workman to raise his wheelbarrow to the height of the bin. He could also put a wedge under the wheel to stop the wheelbarrow rolling back down again.

▶ A true diamond—as it exists in the ground—is a double four-sided pyramid. To make a jewel, the diamond must be cut so that it has many more points and sides. The light then bounces around the space inside the diamond, which makes the diamond sparkle.

▶ Roofs are often shaped like pyramids or wedges, so rain and snow will quickly slide off. What shape is your roof space?

21

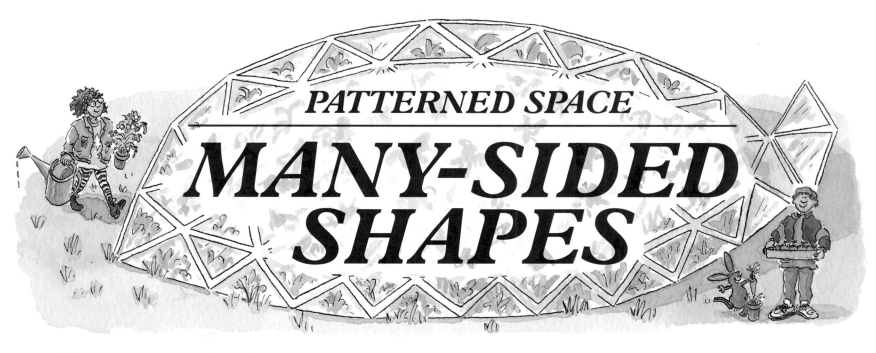

MANY-SIDED SHAPES

In nature you'll find thousands of many-sided shapes—shapes that make beautifully patterned spaces. Look at honeycombs! Look at tortoises' shells, pineapples, crystals, and snowflakes!

We can also make our own many-sided shapes using squares, triangles, rectangles, and hexagons. These many-sided shapes are called polyhedrons. If all the sides are exactly the same shape and size they are known as regular polyhedrons. But most polyhedrons are made up of a mixture of shapes.

22

*pentahedron
(five-sided)*

*hexahedron
(six-sided)*

*icosahedron
(twenty-sided)*

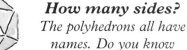

How many sides?
The polyhedrons all have names. Do you know them? The words all come from ancient Greek. Polyhedron itself simply means "many sides" ("poly" means many and "hedron" means side). Each name is taken from the number of its sides.

*heptahedron
(seven-sided)*

*dodecahedron
(twelve-sided)*

*octahedron
(eight-sided)*

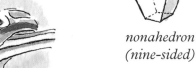
*nonahedron
(nine-sided)*

▶ The Epcot Center is a famous giant polyhedron. It may look like a sphere, but in fact it is made up of nearly 1,000 triangular panels that combine to make hexagons.

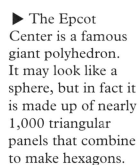

Well done! As a champion of shapes and spaces you can stand on the many-sided podium. Now follow the hare to the ferris wheel and beyond! Count the number of sides on each polyhedron you find.

▲ Why do bees use hexagons to make their honeycombs? Maybe it's because six sides fit together snugly and make a strong shape. This allows the bees to store lots of honey in a small space.

▼ The outside shape of a snowflake is always six-sided. Each flake is a unique and beautifully formed crystal of frozen water. Among all the millions and millions that fall in a snowstorm, no two snowflakes ever have exactly the same shape.

A prickly question
How many sides does a pineapple have? You will have to find one and count for yourself! The skin is made up of a set of little, bumpy disks. But be careful! Each of these disks has its own prickly spine.

Slowly does it!
Its strong shell provides a tortoise with a many-sided space to live in. The shell is made up of a number of plates, which grow larger throughout the tortoise's long life.

SPHERICAL SPACE
BALL SHAPES

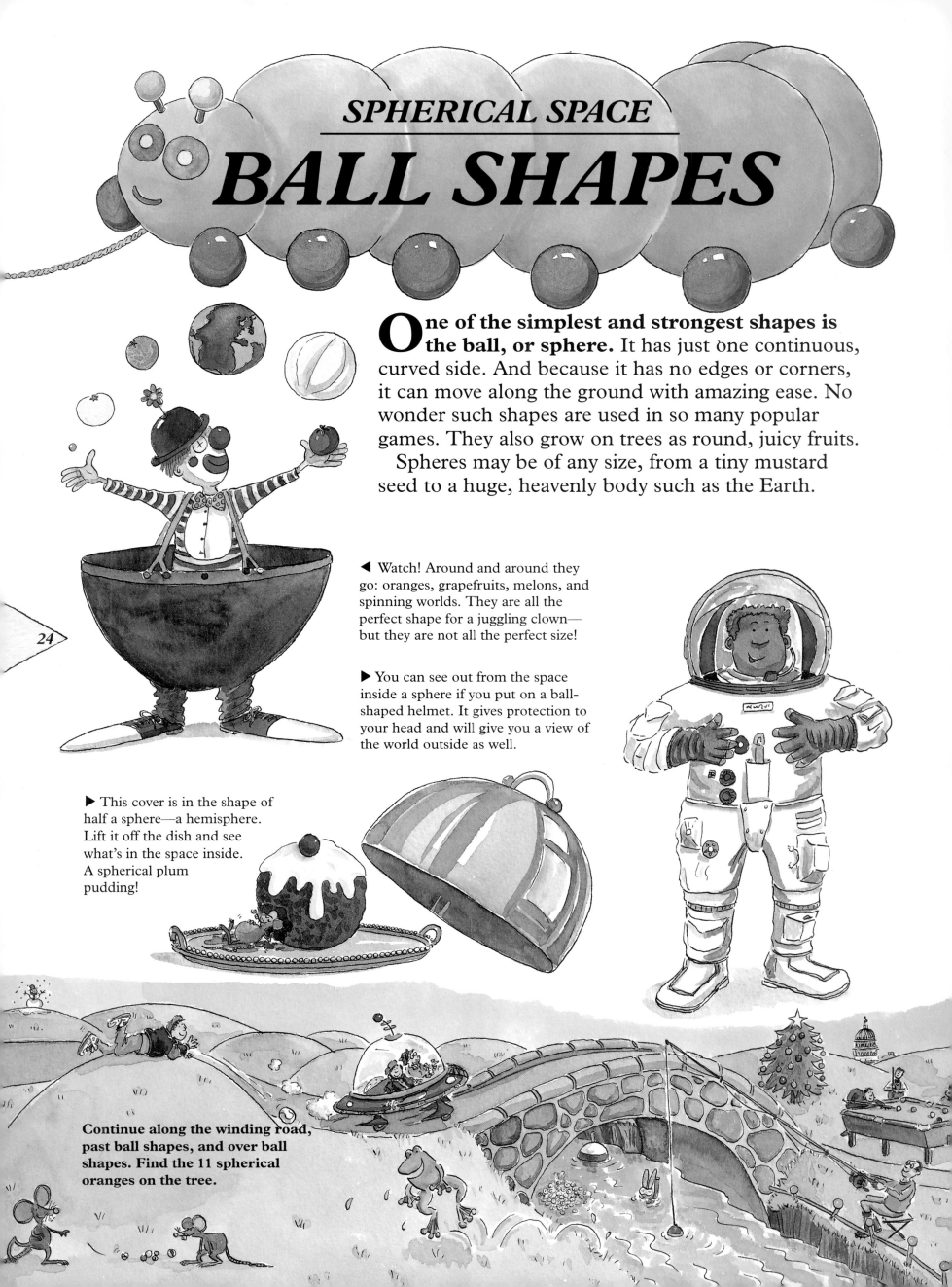

One of the simplest and strongest shapes is the ball, or sphere. It has just one continuous, curved side. And because it has no edges or corners, it can move along the ground with amazing ease. No wonder such shapes are used in so many popular games. They also grow on trees as round, juicy fruits.

Spheres may be of any size, from a tiny mustard seed to a huge, heavenly body such as the Earth.

◄ Watch! Around and around they go: oranges, grapefruits, melons, and spinning worlds. They are all the perfect shape for a juggling clown—but they are not all the perfect size!

► You can see out from the space inside a sphere if you put on a ball-shaped helmet. It gives protection to your head and will give you a view of the world outside as well.

► This cover is in the shape of half a sphere—a hemisphere. Lift it off the dish and see what's in the space inside. A spherical plum pudding!

24

Continue along the winding road, past ball shapes, and over ball shapes. Find the 11 spherical oranges on the tree.

◄ Measure a round watermelon. First find its *circumference*: this is the measurement all the way around its middle. Next find its *diameter*: cut it in half and measure across its width. Then find its *radius*: measure from its center to its edge.

circumference

diameter

radius

Blowing bubbles

Blow very, very gently. See how the thin film of soap carries your breath away in beautiful bubbles. The soap encloses your breath into the neatest, tightest space possible—a sphere!

▲ Basketballs, tennis balls, baseballs, soccer balls, marbles, round balls of every kind—spheres are perfect for playing sports and games. In each ball game, finding the right spaces—nets, goals, gaps, and holes—is what matters most!

ROUNDED SPACE
OVAL SHAPES

Imagine an egg—the beautiful rounded space created and laid by a bird. It is a true oval shape. Like a sphere, it has one single surface without corners or edges. But unlike a sphere, it has two ends.

Some oval shapes look as though they have been squeezed at one end. We call this a pear shape, but there are other things that have this shape besides pears—such as some light bulbs and balloons. Even eggplants are pear-shaped!

▼ You can make oval shapes and pear shapes of your own. Squeeze a round balloon in the middle and what does it become? An oval shape. Squeeze an oval balloon at one end and what do you get? A pear shape.

▶ Hot-air balloons can be many shapes, but this is the most typical. The balloon needs a narrow space at its bottom end for the flame to heat the air, and a big, wide space above to hold this warm air. Warm air rises, so it lifts the balloon and makes it float away.

Are you getting hungry? Look out for some oval-shaped food that you can eat along the way!

▶ How does an egg get its shape? Just before eggs are laid, their shells are soft. The shells only harden on contact with air. With the first push, out comes the rounded end, and with a squeeze, the pointed end follows.

▼ What's inside the newly laid egg? The perfect space for a chick to grow before it hatches.

Are you an egg-head?

Brainy people are sometimes called "egg-heads," but everyone's head is egg-shaped. If you want to paint a portrait of someone, start by drawing an egg shape. Notice how the eyes come about halfway down. You can add hair and features after you get the basic shape right.

27

Can you see a pair of pears?

A pear shape is a natural shape for hanging. Think of a pear hanging from a tree, or the drip on the end of a leaf after a rain storm. Light bulbs sometimes hang down, but can also point up. Light bulbs are made by blowing a bubble of hot glass—like blowing up a pear-shaped balloon.

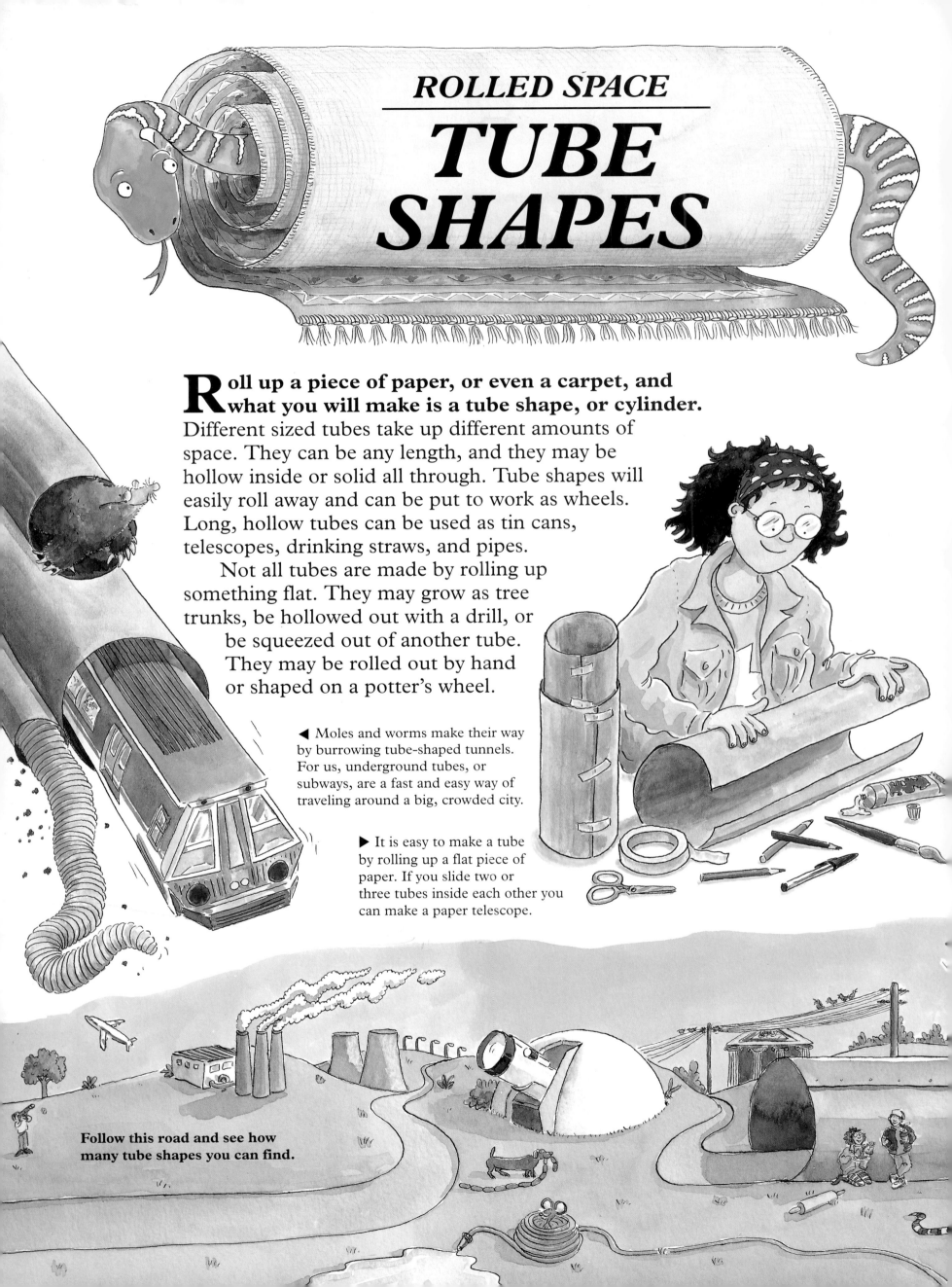

ROLLED SPACE
TUBE SHAPES

Roll up a piece of paper, or even a carpet, and what you will make is a tube shape, or cylinder. Different sized tubes take up different amounts of space. They can be any length, and they may be hollow inside or solid all through. Tube shapes will easily roll away and can be put to work as wheels. Long, hollow tubes can be used as tin cans, telescopes, drinking straws, and pipes.

Not all tubes are made by rolling up something flat. They may grow as tree trunks, be hollowed out with a drill, or be squeezed out of another tube. They may be rolled out by hand or shaped on a potter's wheel.

◀ Moles and worms make their way by burrowing tube-shaped tunnels. For us, underground tubes, or subways, are a fast and easy way of traveling around a big, crowded city.

▶ It is easy to make a tube by rolling up a flat piece of paper. If you slide two or three tubes inside each other you can make a paper telescope.

Follow this road and see how many tube shapes you can find.

▶ When you sip a drink through a straw, it goes into your mouth. Then it travels down through a tube called the gullet into your stomach.

Squeezing a tube
Squeeze a tube of toothpaste and out through the tube-shaped nozzle comes another long, soft tube. You can also squeeze confectioner's sugar through a nozzle like this and make beautiful, wormlike patterns to decorate a cake.

▲ The ancient Egyptians found that tree trunks made very useful rollers for helping to move heavy blocks of stone. Maybe it was this kind of roller that led to the invention of the wheel. Many wheels today have a tube inside that is inflated with air. They are shaped like doughnuts and make the wheels travel over the ground more smoothly.

▶ Tube-shaped containers— cans, jars, pots, pans, bins, and bottles—are easy to make and handy to use. They have no awkward corners and are very strong. On a supermarket shelf they may look neat, but when they are empty they become trash that takes up a lot of space!

CONE SHAPES

Sharpen one end of a tube to a point and you will make a cone. A cone creates a neat space that fits comfortably into your hand—a perfect shape for holding ice cream! A cone can also fit neatly over your head as a hat, or even over a house as a roof. It's a shape you can find in nature—as a pine tree in a forest or a shell on the seashore. It's also a great shape for shooting into space!

30

▼ Make a cone of your own. Just cut out a segment from a circle of card like this. Then roll up the rest of the circle around the center and stick down the outer edge with tape. That's all it takes!

▲ Ten, nine, eight, seven, six, five, four, three, two, one . . . BLAST OFF! Rockets usually have a cone on the top. The cone is streamlined. That means air can rush by easily, which helps the rocket as it speeds into space.

There are 12 people (and a puppet!) along the road wearing cone-shaped hats. Can you find them?

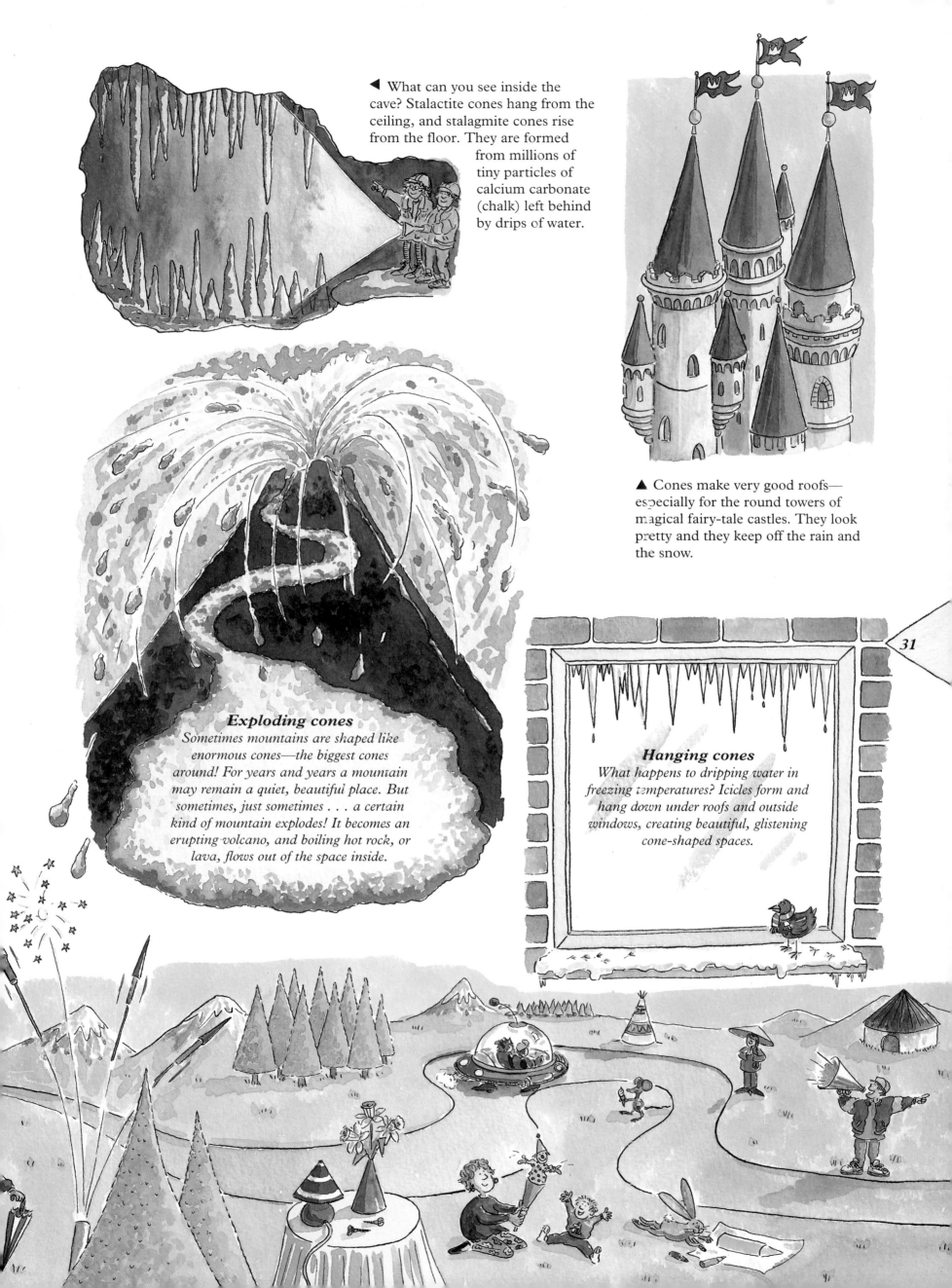

◄ What can you see inside the cave? Stalactite cones hang from the ceiling, and stalagmite cones rise from the floor. They are formed from millions of tiny particles of calcium carbonate (chalk) left behind by drips of water.

▲ Cones make very good roofs—especially for the round towers of magical fairy-tale castles. They look pretty and they keep off the rain and the snow.

Exploding cones
Sometimes mountains are shaped like enormous cones—the biggest cones around! For years and years a mountain may remain a quiet, beautiful place. But sometimes, just sometimes . . . a certain kind of mountain explodes! It becomes an erupting volcano, and boiling hot rock, or lava, flows out of the space inside.

Hanging cones
What happens to dripping water in freezing temperatures? Icicles form and hang down under roofs and outside windows, creating beautiful, glistening cone-shaped spaces.

EVERYDAY SPACE
IRREGULAR SHAPES

Look around you! Many spaces you see every day are irregular in shape. Some of them are different regular shapes joined together, such as buildings with all their bricks, tiles, pipes, roofs, and windows. Other irregular spaces may be regular shapes that are now out of shape, such as a pear with a bite out of it. But many spaces are irregular in shape to start with—just look at rocks, most trees, and animals. It's not always easy to find words to describe such shapes. You try!

◀ Build your own locomotive. Find how many different regular shapes you can use to make it: box shapes, ball shapes, cone shapes, tube shapes. Which shapes do you need most?

Watch out for strange shapes! Which one catches your attention most as you head along the road? What regular shape does it most remind you of?

32

How to make a space irregular

First find a nice, regular shape—maybe a paper hat or an ice-cream cone. Now squeeze it, squash it, roll it up, fold it, crumple it, cut it, slice it, blow into it, bite it, heat it, wash it, dissolve it—or simply leave it to rot away. See how you have changed the space it uses.

▲ Most vegetables refuse to grow in regular shapes. But they still look good, taste good, and fit nicely in the empty spaces of your stomach.

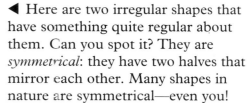

Shapes in the sky

If you want to see really beautiful irregular shapes, look up. In the wide open space of the sky, clouds constantly change shape, forming all kinds of irregular spaces. Sometimes they look like other things. If you use your imagination, you might see animals, castles, ships, and islands in the sky.

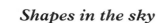

◀ Here are two irregular shapes that have something quite regular about them. Can you spot it? They are *symmetrical*: they have two halves that mirror each other. Many shapes in nature are symmetrical—even you!

SUBSTANCE

Almost every shape and space in the world is filled with some substance or other. The substance may be solid, as in rocks and ice. It may be liquid, as in rivers, seas, and oceans. It may be gas, as in the air we breathe. Completely empty spaces are called vacuums and rarely exist in the world.

Spaces fill and empty, sometimes with our help. When we go shopping, we empty the shelves and fill a shopping basket. Then we empty the shopping basket and fill our car. When we get home, we empty our car and fill our cupboards.

▼ Inside one space is another. And inside that is another . . . and another. Each time the space inside becomes smaller—just like taking apart the famous Russian wooden dolls.

▲ Fill up your stomach! Fill up your clothes! Fill up your suitcases! What happens? You can hardly move!

Look! What solid spaces can you see along the road? Find some spaces filled with gas. Take care not to get soaked by the spaces filled with liquid! Where are they?

34

Water, water, everywhere?

What fills most of the space in all living things? Believe it or not, it is water! About three-quarters of the human body consists of water. And plants contain plenty of water too—if we don't water our gardens, nothing will grow!

Water magic

When water freezes, what happens? The liquid swells and turns into solid ice. When it boils, what happens? The liquid turns into steam and disappears into the air as invisible vapor. Water fills space in all sorts of ways!

▲ Blow up a balloon and see for yourself how you can turn one little flabby space into a great big, shaped one, filled with air. Or buy a balloon filled with helium gas, which is lighter than air, and hold on tight or it will take off!

MEASURING SPACE
VOLUME

The amount of space filling any shape is its volume. The volume of a hollow shape is easy to measure—see how much water (or sand) is needed to fill it. With most regular shapes, you can also use mathematical calculations to figure out their volumes. First measure the lengths of a shape's sides, then follow a formula to find out exactly how many small cubes—cubic inches or feet, for example—would fit inside. To measure the volume of a solid irregular shape you can plunge the shape under water, then measure the rise in the level of the water.

◀ What happens when you climb into a bath full to the brim with water? The floor gets wet, of course, because water is pushed out by the space your body uses. If you could collect the spilled water and measure it, you would be measuring the volume of your body. You can measure the volume of anything—regular or irregular—in the same way. A Greek named Archimedes first thought of this idea about 2,250 years ago, and he has been famous for it ever since.

As you travel along the road, look at all the things that take up space. Everything does! And buckets, pots, bottles, and jugs are all used to measure volume.

Compare these volumes

The average adult person has a volume of about 3 cubic feet.

An African bull elephant, the largest animal on land, has a volume of about 180 cubic feet—the same as about 60 people.

A blue whale, the largest living animal, has a volume of about 4,500 cubic feet—the same as about 25 elephants.

All of the oceans and seas combined have a volume of about 300 million cubic miles—the space of billions and billions of blue whales.

The whole world has a volume of about 260 billion cubic miles—in which all the oceans and seas could fit 850 times over.

▼ Volume can be hard to judge. For example, which of the regular shapes shown here do you think has the largest volume? Is it the sphere, the pyramid, the cube, or the cone? Trick question! They all have the same volume.

▼ There are many different ways of thinking about the space inside things. You can measure a drawer by finding out how many shirts will fit into it. You can measure a wading pool by seeing how many bucketfuls of water it takes to fill it.

Same volume, different spaces
To make Jell-O™, just add hot water to the mix in a bowl, then pour the mixture into a shaped Jell-O™ mold. The volume of Jell-O™ remains almost the same, but the shape of the space it occupies changes.

MAPPING SPACE
WHERE?

Where are you? How would you describe exactly where you are right now? Do you know the way to your kitchen or to the nearest store? To reach either of these, you need to know the layout of the space around you, whether in your home or your neighborhood. At some time you have explored this space and remembered it. For places farther away—places that are new to you—you may find it useful to look at a map. By making and using maps, people can explore spaces near and far and really know where they are.

38

▼ People who help make maps—surveyors—have to do a lot of measuring. They use instruments called theodolites to measure the spaces between things, such as buildings and bridges, the lengths of roads, and the heights of hills and mountains.

▼ The X shows where the buried treasure lies. Where should you go to find it—east, west, south, north, or maybe southeast? With a good map you should be able to find your way. The map tells you what to look for, and where to turn left or right.

NORTHWEST   SOUTHEAST

Use the treasure map above to find where the gold is hidden. Go southeast through the forest. Cross the river. Follow the path over the mountain and through the swamp. Take four more paces and start digging!

▶ Can you pinpoint exactly where everything is in your room: your bed, your table, your light switch, your secret hiding places? You can make a map of your room by drawing a grid plan on a piece of paper. Start with the outline. Then imagine you are a fly on the ceiling, and draw what you see below using the grid lines as guides.

▼ Waiter! Where's my soup? It's not easy being a waiter in a crowded restaurant. Waiters must serve food quickly to hungry customers. They develop a kind of map of the restaurant in their minds, so they can take the right food to the right table by the shortest route. We all have maps in our minds. Think of how you would take food from your fridge to the dinner table.

Say cheese!
When you take a photo, you are making a kind of instant map of the world in front of the camera. A photo makes the world look flat and still, but of course it is not. Maps make the world look flat and still as well. But symbols on the map help tell how the world will really look when you follow the map.

OBSERVING SPACE
OUT THERE

Look! Listen! Feel! You need all your senses to observe fully the space and spaces around you. As we move and travel, we can observe more and more. The farther we go—by car, train, plane, or space rocket—the more we see. We can also observe more when we stay in one place, by using scientific instruments that increase the power of our senses. Telescopes, for example, help us to see things so far away that we couldn't see them with our eyes alone, and we could never travel to them.

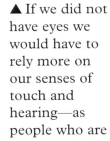

▲ If we did not have eyes we would have to rely more on our senses of touch and hearing—as people who are blind do. Try putting on a blindfold and finding your way around a room that you know. What do you notice more because you cannot see?

◄ Our eyes can see long distances, but with the help of binoculars or a telescope all sorts of new things come clearly into view, whether we are looking at a bird in a tree or up at the moon. Try it and see!

Hurry along now! Look at all the different sorts of transportation. Find a train, a car, a boat, a bicycle, and a skateboard. What else do you see?

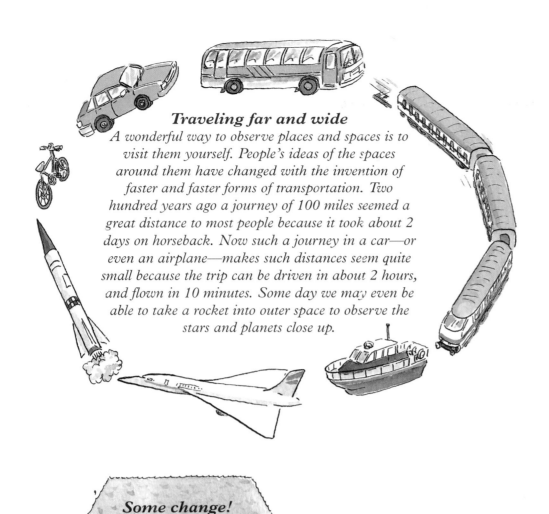

Traveling far and wide

A wonderful way to observe places and spaces is to visit them yourself. People's ideas of the spaces around them have changed with the invention of faster and faster forms of transportation. Two hundred years ago a journey of 100 miles seemed a great distance to most people because it took about 2 days on horseback. Now such a journey in a car—or even an airplane—makes such distances seem quite small because the trip can be driven in about 2 hours, and flown in 10 minutes. Some day we may even be able to take a rocket into outer space to observe the stars and planets close up.

▼ How do we know about the spaces under the oceans? Scientists on boats can measure the depth and shape of the seabed by sending out underwater sound signals and measuring how long the sounds take to bounce back.

Some change!

Five hundred years ago, people had many strange ideas about the space out there. Most people believed that the world was flat, and if they went too far they would fall off the edge! Then, in 1522, Ferdinand Magellan's ship returned home after completing the first journey around the world, proving that the Earth could not be flat.

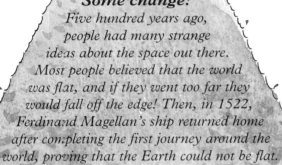

Orion

The Great Bear

◀ Up in the night sky people have picked out different patterns of stars. We call these patterns constellations, and give them names, such as Orion and the Great Bear. Together all the constellations make up a kind of map of the night sky. Astronomers observe the stars and planets in great detail by using powerful telescopes in observatories like the one below.

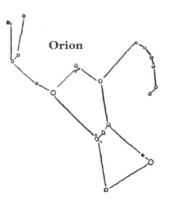

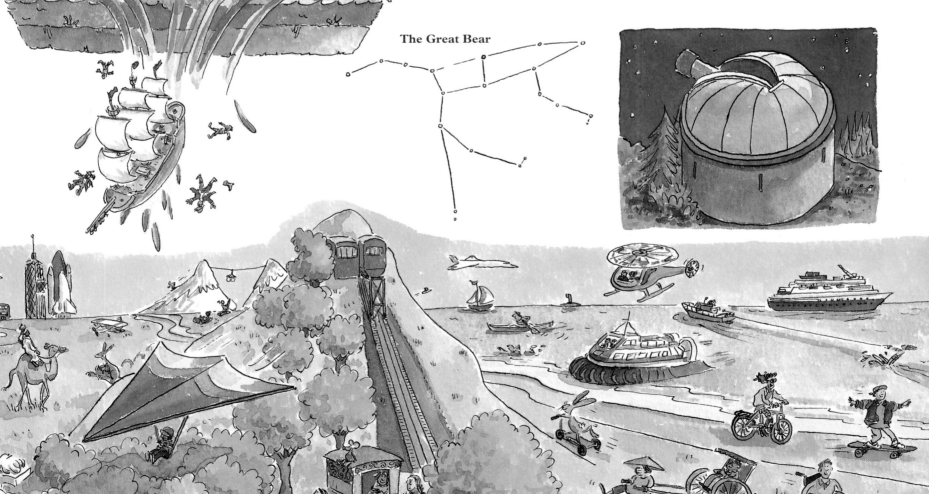

MAGNIFYING SPACE
IN THERE

What goes on in the thinnest, smallest parts and innermost regions of space? If we magnify what's inside any substance we discover an amazing world that we cannot see without help. Scientists do this with the help of microscopes, some of which are extremely powerful.

Look at a ladybug, for example. The more we magnify it, the more we see. Magnification will finally reveal that a ladybug, like everything else, is made up of billions of moving atoms in a space that is very, very different from the one we normally see with our eyes.

▲ The first microscopes were invented more than 300 years ago. They contained glass lenses to magnify the things people wanted to look at very closely. Modern microscopes with lenses can magnify things up to 1,500 times their normal size.

42

▶ An electron microscope can magnify something up to several million times, and produces clear images on a special kind of television screen.

Ladybug— actual size

Ladybug— magnified 10 times

▶ A magnifying lens shows you what tiny insects really look like. Here is a ladybug enlarged to about 10 times its actual size. Each of its spots now looks bigger than one ladybug.

Welcome to the Museum of Magnified Space! Discover the secrets of inner space! Find each of these microscopic things—cells, molecules, and atoms.

Magnified animal cells

Magnified plant cells

MUSEUM OF **MAGNIFIED** SPACE

Carbon Dioxide Molecule

Ladybug Molecule

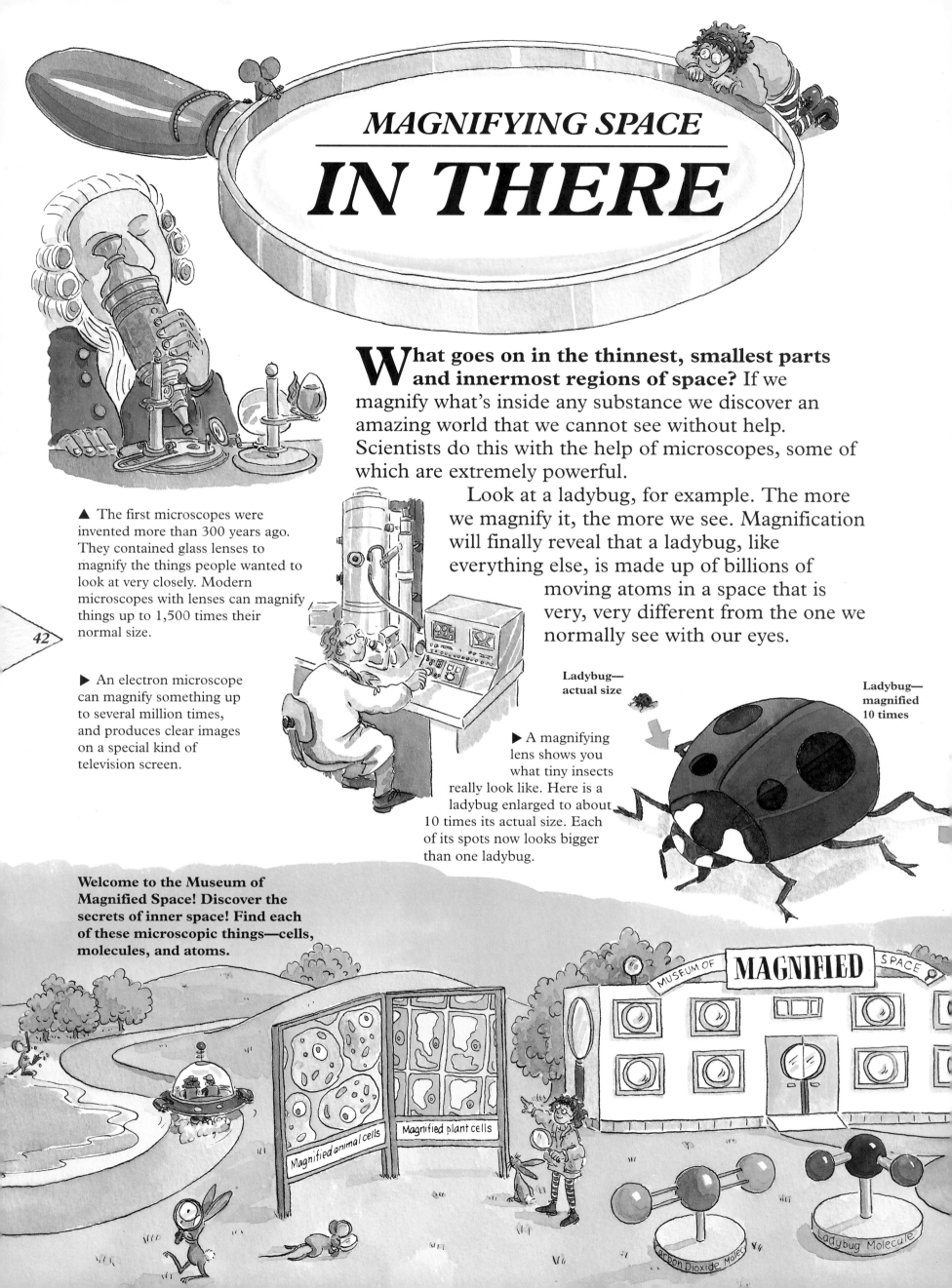

Microscopic life

Scientists were astonished 300 years ago to find strange bodies wriggling in water under their microscopes. They were discovering tiny living things known as microorganisms, which are too small for the eye to see. These include single-celled plants and animals, bacteria, and viruses (which are too small to see with a lens microscope). Some bacteria and viruses cause diseases.

▶ This is an oxygen atom, such as might belong to a ladybug. Atoms are extremely small. This atom has been magnified about 50 million times. But atoms can be divided into even smaller particles. This atom is shown with electrons circling around the nucleus in the middle. A single page of this book is about a million atoms thick.

Ladybug— magnified 50 million times

▶ An electron microscope that magnifies 10 million times reveals that a ladybug cell is made up of many molecules, including molecules of water. A water molecule consists of two atoms of the gas hydrogen and one of the gas oxygen. A single drop of water contains about 3,000 million million million molecules.

Ladybug— magnified 10 million times

▼ A lens microscope can magnify each spot on the ladybug more than 200 times and show that it is made up of hundreds of cells that are each like this. Cells are constantly dividing in two to create new cells.

Ladybug— magnified 200 times

▶ One space that many of us never see is the one inside our own skins. But it is no mystery to doctors. X-ray pictures show doctors the shapes of our bones—and whether any of them are broken.

43

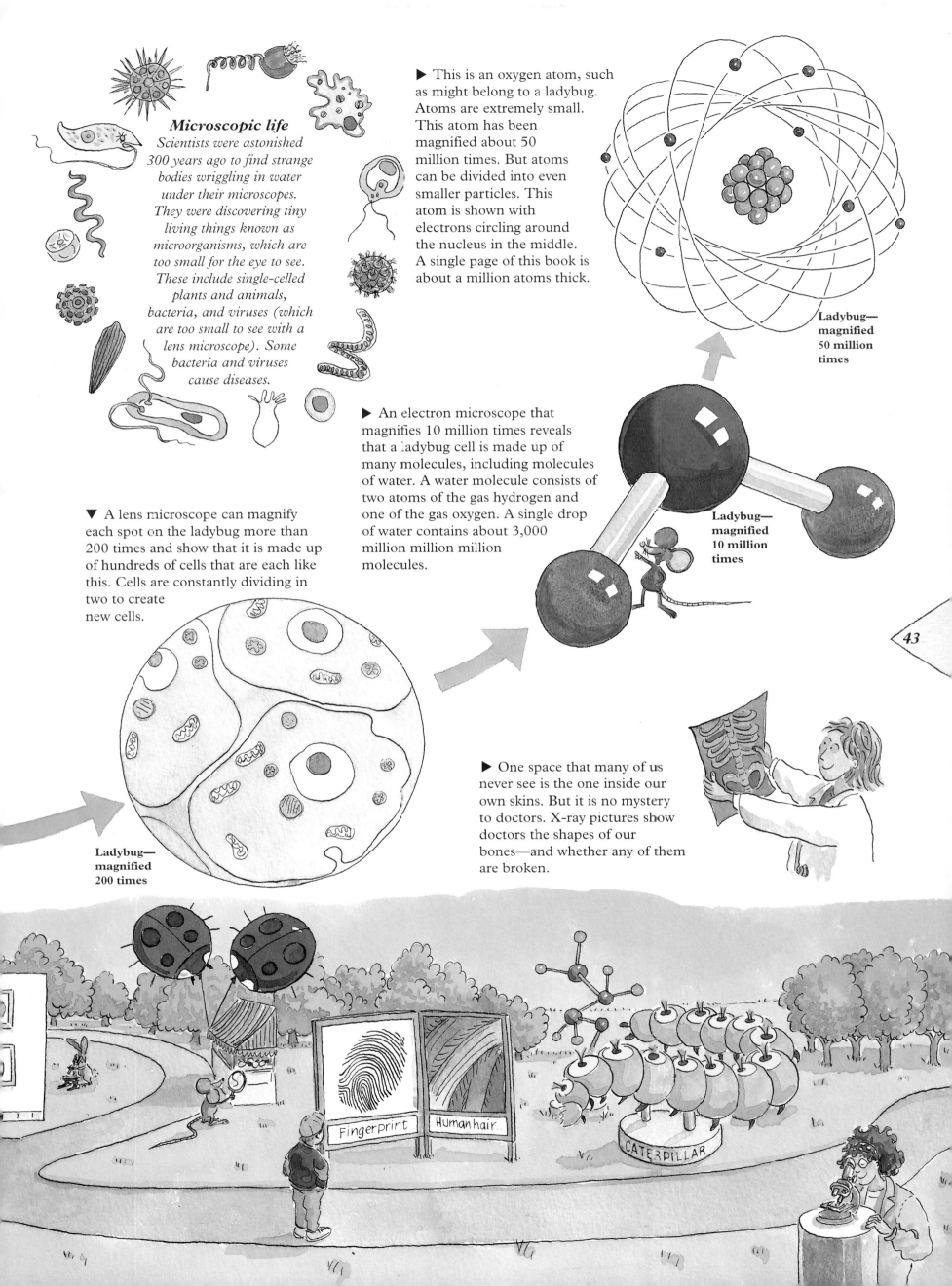

Fingerprint   Human hair

CATERPILLAR

ALL SPACES
EVERYWHERE

Everywhere we turn and everywhere we look—in every shape, substance, or place—there are spaces and space, and more space in between. Our journey has taken us over vast distances and into minuscule places, but now we must stop. We've run out of space in this book!

It's time for you to stand back and think about all you have read and seen. But most of all, it's time for you to start exploring the wonders of space for yourself.

◄ Cones, balls, wedges, boxes, regular shapes, irregular shapes—it takes all sorts of shapes to fill space. Look around you. How many shapes can you now name?

The road keeps going—in and around, up and down, over, under, and through every space you can imagine, even to outer space!

A never-ending story?
This story could go on forever.
Each space is part of another
space, which is part of another one,
which is part of another—from tiny
atoms to the space inside our bodies,
to our homes, our neighborhoods,
our world, our solar system, our
galaxy . . . even the universe!

45

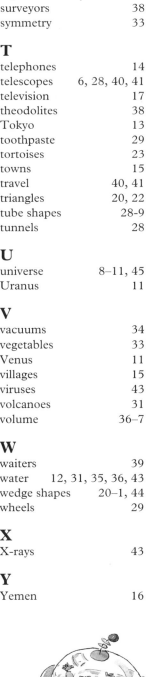

INDEX